Dedication

To those who never came home.

 Published in the United States by Veterans Publishing Systems - Adducent, Inc. Jacksonville, Florida. www.AdducentInc.com.

Printed in the United States of America

ISBN 978-0-9777884-1-5

Library of Congress Control Number: 2009909698

Table of Contents

Acknowledgements

First and always, to my beloved and patient Kathleen, my companion, mentor, friend and editor. Colonel Bill Schneck walked into my office one day with a stack of source material on Gooch's Regiment. This, and a good deal more, was in that stack. Thanks Bill, it has been fascinating and humbling. Mike McNally, noted author and friend, helped make this far better than it would have been. Thank you, Arthur for the last minute ideas that helped make the story live. Last but not least, my thanks to Lieutenant Colonel Zumwalt, U.S.M.C. Retired, who guided me through the process of getting this story published.

A note on the use of quotes: An effort has been made to normalize 18th century punctuation, spelling and capitalization to ease reading and comprehension.

Preface

The struggle for control of the Caribbean basin began in the late 16th century. Spanish control of the gold and silver mines of South and Central America was contested by Holland, England and France. Control of the sugar trade, the raw material for rum was almost as valuable and equally contested by the major European Powers of the time. The slave trade, a hideous, but critical requirement for the sugar trade was also a valuable commercial activity. All these contributed to make the region a source of commercial and military competition. North American colonists had been involved in trade, both legal and illegal, in the Caribbean since the founding of those colonies. The Colonial military contribution had always been in the form of privateers sanctioned by the British Government in time of war, and pirates when government sanctions were not available. The British expeditionary operations of 1740-1742 are another chapter in the ongoing struggle of national control of the Caribbean with the unique American contribution of soldiers. This had not been attempted before and was never attempted again.

Introduction

The War of Jenkin's Ear began on 22 October 1739, with a declaration of war by Britain on Spain. It was another event in a multi-national Caribbean trade dispute that dated to the 16th century. The war's name comes from an incident that occurred on 9 April 1731 when Captain Juan Francisco, a Spanish "Guarda Costa" intercepted the British brig Rebecca and cut off the ear of one Captain Jenkin's, causing belated outrage in England and the Parliament, while giving the British Government the excuse for a declaration of war eight years later.

In October of 1739, Vice Admiral Vernon arrived at Port Royal Jamaica with a fleet of nine ships with orders to make an attempt on Portobello, an important Spanish port on the Atlantic side of the Isthmus of Panama. From Jamaica, Vice Admiral Vernon sailed on to Portobello after having added approximately 200 soldiers to his 2000 sailors. Arriving at Portobello on 21 November 1739, the town surrendered after a battle of perhaps no more than three hours. The loot was not as plentiful as expected and little was achieved in the strategic sense by his immediate departure, which allowed the Spanish to reoccupy the city. However, the psychological effect on the British people, military establishment and, perhaps on Vice Admiral Vernon, was considerable. The belief that the Spanish Caribbean ports were rich and vulnerable became the accepted view. In February of 1740, Vernon sailed to Cartagena, which was, at the time, the best fortified and one of the largest Spanish towns in the Caribbean. He bombarded it, but found it too strong to attack with his small fleet. He captured Fort Chagres at the mouth of the Chagres River on 24 March, left it unoccupied, and sailed back to Jamaica to await reinforcements.

The perceived vulnerability of the Spanish led to the organization of an expedition, under the joint command of Vice Admiral Vernon and Lord Cathcart, to the West Indies to target whichever port the senior leaders of the expedition deemed best. The expedition led to the raising of six new Marine Regiments in England (Wolfe's, Douglas', Lowther's, Robinson's, Wynard's and Morton's), with cadre from existing regiments, as well as the inclusion of Cavendish's 34th Regiment of Foot and Bland's 36th Regiment of Foot. This made for a land force of approximately 6,500 men and was seen as an insufficient number for the expedition. It was not possible to raise any more men in Great Britain due to concerns over the war in Europe. The Crown began to look elsewhere for soldiers.

A new regiment was raised in the North American colonies to participate in the expedition. The regiment was to be commanded by a former Lieutenant Governor of Virginia, and called "Major General Alexander Spotswood's Regiment of Foot." The Regiment was unique in the way it demonstrated, and led to the publicizing of, the emerging differences between the American and European subjects of the British Crown. We see from accounts in colonial newspapers that the North American Colonists were aware of the conditions of service and subsequent losses experienced by the Regiment. One can only speculate as to the effect of those accounts on the readers. What we can document however is a new reluctance to serve under British Officers and in accordance with the British Mutiny Act during the Great War for Empire that began thirteen years after Gooch's Regiment was disbanded.

The following is the first comprehensive overview of Gooch's Regiment of Foot's entire period of service. The Regiment was unique in its composition, organization, staffing and size. It brought together companies from nine of the North American Colonies under American leadership in a manner not attempted before, or since, for service as part of an expeditionary force deployed to the Caribbean. The Regiment was raised as a Foot Regiment but served as Marines, a common practice of the time. **They were America's first Marines.**

1

Organization and Uniform

The Regiment was approved by Parliament on 25 April 1740 to be organized into four Battalions and a total of 30 companies. The Regimental staff and Battalion Commanders and Staff all came from the British Army. One of two Lieutenants in each of the Companies was appointed by the Crown and one Sergeant from each company came from Phillip's 40th Regiment of Foot and the four New York Independent Companies.

> *"The lieutenants were appointed in England, at the recommendation of Lord Cathcart, who commanded all the land forces on the intended expedition; and he chose for this service young gentlemen of family, chiefly North Britons, who had learned the rudiments of the military art in Holland and other foreign services, and consequently were the better qualified to discipline a new raised regiment."*
> **Tobias Smollett on Gooch's Regiment in**

"An Account of the Expedition against Carthagena in the West Indies"

> *"Orders should likewise be sent without loss of time to Col. Phillip's Regiment in Nova Scotia and to the four Independent Companies at New York to furnish these troops with a draft of 160 private men, old soldiers to serve as Sergeants and Corporals."* **Calendar of State Papers Colonial, December 1739**

The Regiment had, in addition to the Commander, a Second in Command, Chaplain and Surgeon. Each Battalion would have a Lieutenant Colonel as Commander, a Major as Second in Command, an Adjutant, and a Surgeons' Mate. Each company of 100 men would have a Captain, two Lieutenants, and one Ensign as well as four Sergeants, four Corporals, and two Drummers. Total Regimental numbers were to be 13 Field Grade Officers, 30 Company Commanders, 60 Lieutenants, 30 Ensigns, 120 Sergeants, 120 Corporals and 3000 men for a total of 3373. The actual numbers raised exceeded those approved by Parliament with 36 Companies and approximately 3600 men raised.

> *"It is His Majesty's intention, to give all proper encouragement to the new levies,*

> *by ordering them to be supplied with arms, and a proper clothing, and to be paid by His Majesty; with an assurance of their coming in for their share of any booty, that may be taken from the enemy, and to be sent back to their respective habitations, when the service shall be over, unless any of them shall desire to settle themselves elsewhere."* **Duke of Newcastle to Governor Thomas on the Expedition to Spain, West Indies, 5 January 1740**

Soldiers raised for Gooch's Regiment received the same rank and pay as other "regular" soldiers, but unlike their regular counterparts, they were entitled to keep their arms and uniforms once their service ended. The date pay commenced for the NCOs and officers was from the date of commission, but their seniority was based on date they completed their requisite levy's of troops, as rank was awarded as a result of troops raised, not military experience. Unlike other Officers in the British Establishment, Officers from North America commissioned for service in Gooch's Regiment were not entitled to retire on half pay once their service was complete.

The "Cloathing Book of 1742", commissioned by the Duke of Cumberland for the King, identifies the 43rd Regiment of Foot as having been "raised in America" and commanded by Major General Spotswood in 1740 and later Colonel Gooch. It also shows Gooch's Regimental uniform as consisting of a red coat with green cuffs and lapels, red waistcoat and red breeches. The coat lining is depicted as being light brown, perhaps to indicate the unbleached linen characteristic of uniform coat linings for units on tropical service. The usual black three-cornered hat with white trim, buff leather belt and buff leather strapped black cartridge box, gaiters and black shoes are depicted as well.

One of the commonly cited sources for Gooch's regimental uniform is Boyse's two volume account of the events around the globe during the period of the war, from 1739-1745, which was published in 1747. In his one paragraph description of Gooch's Regiment, he is wrong about the number of troops, where they were mustered and how they were organized. This certainly calls into question his description of the Regiment's uniform as being *" red camblett coats, brown linen waistcoats and two pair of canvas trousers each man."* Another consideration was the promise made by the Crown that Gooch's Regiment *"shall be provided for in every other aspect as our British troops."* The uniform described by Boyse clearly does not meet that standard.

Boyse is also consistently unreliable in his description of the terrain, fortifications and events at Cartagena. However, Boyse is in concurrence with other contemporary sources in his, albeit brief, description of the events in Cuba in 1741. In Boyse's defense, it is possible that his uniform description is more accurate for soldiers posted to the Caribbean when his account was published five years after Gooch's Regiment had been disbanded.

2

The Men of the Regiment

> *"The King intends that the troops to be raised in America, should be commanded by Col. Spotswood, the whole to be however, after they shall have joined the regular Troops, under the Command of My Lord Cathcart, General and Commander in Chief of all His Majesty's Troops sent or raised in America."* **Duke of Newcastle to Governor Thomas on the Expedition to Spain, West Indies, 5 January 1740**

Major General Spotswood had been a Lieutenant Colonel in the Earl of Derby's 10th Regiment of Foot. He served under Marlborough and was wounded at the battle of Blenheim in 1704. He was an experienced, well respected, but old, soldier. He had originally proposed the idea of an American Regiment for service against the Spanish in the 1720's. Major General Spotswood was also appointed Deputy Land Force Commander and

Quartermaster General of the expedition. His Executive Officer was to be Lieutenant Colonel Henry Cope, the Governor of Newfoundland.

> *"As an inducement to your people, to engage in the intended expedition, it may not be amiss to remark to them the very prudent care which His Majesty has taken of their interests; by appointing an old experienced Officer (who has resided among them 25 years, and has his estate fixed in this part of the world) to have the particular Command of the North American Forces: whereby they may assuredly expect, that he will be a strenuous stickler in their behalf to have justice done them, in their due share of the booty."* **Major General Alexander Spotswood to Governor Wanton of Rhode Island, 3 April 1740**

Major General Spotswood received his commission to raise a regiment from the King in March, before the funds for the Regiment had even been approved. Not long after, he raised the 400 men for the Virginia Companies and had assembled them at Williamsburg to begin training for the coming expedition. Major General Spotswood died in Annapolis on 7 June 1740 and the

Lieutenant Governor of Virginia, William Gooch, replaced him as Colonel of the Regiment. Colonel Gooch, like General Spotswood, had served under Marlborough as a Captain in Godfrey's 16th Regiment of Foot. He retained the position of Quartermaster General for the expedition, at least initially, but not that of Deputy Land Force Commander.

> *"The General hearing and finding I was better received at Jamaica than himself, and being proud and stingy, tried all ways he could think of to let the people know, that though I was a Governor, yet he was the greater man, and indeed some orders he might have been spared, however, I submitted, and was all obedience, resolving that if military subordination was carried to slavish subjection he should not tire of me."* **Colonel Gooch on Major General Wentworth**

The council of war was the military decision making body of the 18th century. They were held prior to any decision or major operation by the land, and or, sea force. The signatures of some, if not all, of the other Regimental Commanders are found on the resolutions of numerous land force councils of war. There is only one example of the Commander of Gooch's Regiment signing any of the resolutions from the expedition's councils of war. It occurred when the Regiment was under the command of Colonel Cope on 30 March 1742 at Portobello.

The land force for the expedition was divided into three brigades, with the battalions of Gooch's Regiment divided between all three. This left Colonel Gooch with no consolidated Regiment to command. This being the case, one wonders how much of a command function the Regimental Commander and his staff served.

The following is a roster of known company commanders in Gooch's Regiment. The information displayed below was gathered from a variety of sources, but predominantly existing Regimental Returns. What it shows is an absence of consistent colony/battalion affiliation over the period of the Regiment's existence. This is counter to the references in the literature concerning companies from specific colonies assigned to specific battalions in the Regiment.

The popularly assumed battalion/colony associations are indicated at the top of each column. Each company commander's last name has his known colony of origin, and any additional battalion associations. If he was known to have died in service with the Regiment, his name is italicized and followed by a "D". There are officers affiliated with each Battalion, but not known to have been company commanders. There are also known company commanders with no known Battalion affiliation. Captains in these two categories are found below as well.

1st Bn. (MA)

Goff (MA)(3rd Bn.)(D)
Prescott(MA)(3rd Bn.)
Stewart (MA)(3rd Bn.)(D)
Phillips (E)(MA)
Dunn (RI)(D)
Hepburn (SC)
Innis (NC)
Pepper
Walker (E?) (VA)(D)

2nd Bn. (NJ/NY/RI)

Clarke (PA)(4th Bn.)(D)
Provost (NY)(D)
Thomas (NJ)
Hopkins (RI) (4th Bn.)
Cosby(NY)(4th Bn.)
Hooper (NJ)(D)
Farmar (NJ)
Stevens (NY)(D)

3rd Bn. (PA/DE)

Winslow (MA)
Stillman (1st Bn.)
Gibson (D)
Bishop (PA)
Phillips (D)
Newberry (CT)(D)
Stafford (E)

4th Bn. (MD/VA/NC)

McKnight (PA)
Lowery (PA)
Freame (PA)(D)
Washington (VA) (1st Bn.)
Fitzhugh (VA)
Bushrod (VA)(1st Bn.)
Addison (MD)
Lloyd (MD)(lst Bn.)
Colerain (NC)(2nd Bn.)
Gordon (SC)(PA)
Mercer (VA)

4th Bn. (MD/VA/NC)
continued

Downey
Dalrymple
Gillespie (D)
Thynne (PA)
Bishop (PA)
Jenkins (PA)(D)
Holton (NC) (3rd Bn.)
Pratt (NC)
Milburn (MD) (1st Bn.)

Unaffiliated Company Commanders: Anderson, Martin, Boswell, Cooper, Henderson

Captains not identified as Commanders but with Battalion affiliations: Halton (1st Bn.), *Caylor (2nd Bn.)(D), Erskine (2nd Bn.)(D),*

The relatively large number of commanders with no colonial affiliation could very well mean that companies in Gooch's Regiment were commanded by European and not solely American officers, perhaps several times, throughout the life of the Regiment. There are four known examples of European commanders. Captain Hepburn, who was identified as Scottish (indicated with "SC" and in bold), Captain Stafford (indicated with an

"E" and in bold) and Captain Rowland Phillips, an Englishman, (indicated with an "E" and in bold) who commanded one of the Massachusetts Companies. Captain Gordon, a Scotsman (indicated with "SC" and in bold), raised a company in Pennsylvania. Another possible example is Captain Charles Walker, possibly an Englishman (indicated with "E"? and in bold), who commanded one of the Virginia Companies until his death. He was replaced by one of the promoted Ensigns from Virginia, Captain Fitzhugh. There was a mutual antipathy, evident in the literature since the 17th century that typified relations between American and European soldiers and officers. Replacing American commanders with Europeans could very well have had an adverse effect on unit morale and effectiveness.

A note on the use of the term "European" is in order. This word was used by Vice Admiral Vernon in his correspondence to differentiate the men and officers from Britain and the American Colonists. According to Vice Admiral Vernon, it was used by the Americans as well. As a rule, language does not waste terms on distinctions that do not exist. It is apparent that by 1740, there was a perceived difference between British subjects living in North America and those living in Britain. As we know, over time this difference would only become greater.

> *"As for the American troops, they were in general many degrees worse, but the*

> *Officers in particular. Who were composed of blacksmiths, tailors, shoemakers, and all the banditti that country affords, insomuch, that the other parts of the army held them in scorn."*
> **Charles Knowles, "An Account of the Expedition to Carthagena"**

It is reasonable to think that, typical of 18th century regiments, the enlisted men in Gooch's Regiment came from the lower strata of society. In that regard, they were no different from other regiments in the British Establishment. There were no doubt a multitude of motivations to join and it is certain there were those who looked forward to adventure and a share of booty. The Government in England and Vice Admiral Vernon seemed to think, for no apparent reason that the Americans were looking for new land to settle.

Colonies made an effort to provide incentives to enlist. In Maryland, those who enlisted were exempted from arrest for debt. Enlistment bounties were provided by the colonies of Massachusetts, Virginia, Rhode Island and Maryland. In Virginia, indentured servants were permitted to enlist. Where incentives failed, other means were used. The speed with which the Virginia Companies were raised may be due, in part, to a law allowing the drafting of men who met the physical qualifications but lacked an occupation or property.

> *"That it shall and may be lawful, to and for the justices of the peace of every county within this colony, or any three, or more of them, at any time before the first day of December next, within their several and respective counties, to raise and levy such able-bodied men as do not follow or exercise any lawful calling or employment, or have not some other lawful and sufficient support and maintenance, to serve his majesty, as soldiers, in the present war, against the Spaniards, in America."*
> **Laws of Virginia, May 1740**

There are a few clues about the ethnic composition of the Regiment. We know that after raising the regiment no effort was made to recruit in Virginia, Maryland or North Carolina due to Major General Wentworth's perception that he would only find Irish, Catholics or English convicts as recruits in those Colonies. From the record, we have no way of knowing if or how many African Americans actually served in Gooch's Regiment. There was one 50-year-old African American who deserted from the Regiment in September 1740, which indicates the possibility of others. We do have some insight about Native Americans from Massachusetts serving in the Regiment.

> *"Since my last I think I have persuaded a fifth company to proceed without commissions or arms. 30 odd of the company are Indians & very likely men... They are the King's natural born subjects, bred up in the English manner, are good shotsmen, and their having black hair & tawny faces don't at all disable them from being good soldiers. And I have the honor to agree in your sentiments that it will contribute better to the service to have them mixed with English than to be in a body by themselves. I shall give Captain Winslow (who has 30 of them) the kind caution you mention, that his white men should not insult or use them ill."*
> **Governor of Massachusetts and New Hampshire Jonathan Belcher to Colonel William Blakeney, 15 September 1740**

From the following, we see desertion was, as in all 18th century regiments, a problem from the inception of Gooch's Regiment.

> *"Deserted from his Majesty's Service out of the American Regiment of Foot, commanded by Col. William Gooch, and*

lately enlisted in West-Jersey, by Lieutenant Anthony Palmer, the two following Soldiers, viz. Thomas Rymer. an English man, fair complexioned, with light short Hair, about 5 foot 7 inches high, well-set, ...Had on when he went away, a homespun lightish mixed waistcoat and breeches with brass buttons. Thomas Fury, a laborer, born in the North of Ireland, about 21 years of age, 5 foot 10 inches high, well-set, fair complexioned, with very fair eye brows, grey Eyes, and much Pockfretten : Had on when he went away, a greyish homespun coat, with brass buttons, the lowermost but one having the top broke off and in other places some off. Linen trousers, and a pair of new shoes." **The Pennsylvania Gazette, July 22. 1742.**

"Deserted then from Capt. Thomas Freame's Company, in the Hon. Col. Gooch's Regiment, viz. Samuel Bickerstaff aged 20 Years, 5 Foot 6 Inches and 3 quarters high without shoes, brown complexion, well-fed, by trade a weaver, had on a light colored cut fustian jacket and breeches, a good hat and cap, &c.

and Englishman." **Franklin's Gazette August 18, 1740**

The following persons being enlisted in his Majesty's Service in Captain John Milbourn's Company of the honorable Col William Gooch's Regiment have deserted Viz : Daniel Rayne born in Kent County about nineteen years of age five feet seven Inches high well set and of a swarthy complexion had on a corduroy coat and other clothes: James Knott born in Queen Ann's County about 20 years of age five feet ten Inches high, very slender and fair complexion had on a very light colored coat striped jacket and leather breeches. William Burgess a black man about fifty years old six feet and an inch high black hair had on a brown cloth coat and other clothes: John Barnes born in Kent County about 18 years old pale complexion five feet ten Inches high, had on a light colored kersey coat and other clothes,....the following persons having enlisted in his Majesty's Service in Captain Thomas Addison's Company of the same Regiment have deserted Viz : John Gwyn aged about 36 years five feet Eight Inches high fair

Complexion with light brown Hair, had on a striped flannel jacket and a plaid jacket: Anthony Irwyn about 43 years old five feet three Inches high black hair of a swarthy complexion, had on a yellowish coat and other clothes: Richard Gaither born in Ann Arundel County about 22 years old five feet six Inches high had on a dark colored kersey or drugget suit of clothes : John Ward born in Ann Arundel County aged about 24 years five feet nine inches high of a brown complexion, thin visage, the bridge of his nose fallen, had on a dark colored suit: John Woodall a shoemaker by trade about 32 years old five feet seven inches high of a swarthy complexion & sickly countenance had on a coarse linen frock : Francis Atkinson about twenty five years of age five feet eight inches high. John O'Bryan a lusty well made man about five feet nine Inches high an Irishman aged about 38 years speaks pretty good English." **By His Excellency Samuel Ogle Esqr Governor and Commander in Chief in and over the Province of Maryland. A Proclamation," 20 September, 1740**

Theatre of Operations

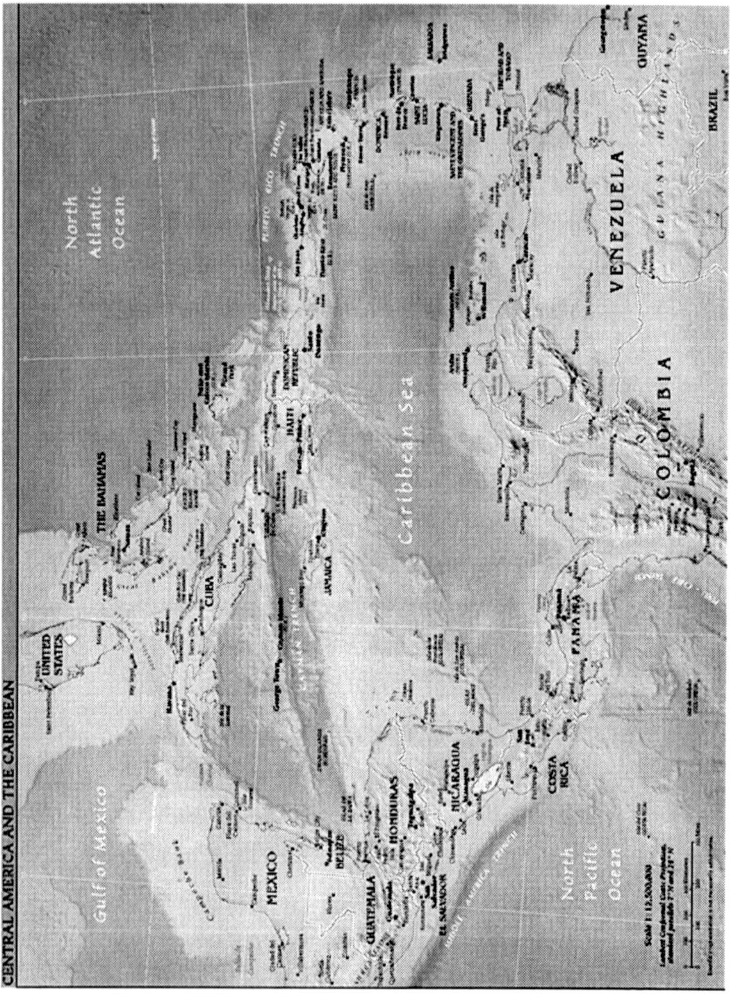

3

Prelude to Action

"As a farther encouragement to such officers and soldiers, as shall engage in the present expedition, you may assure them in Our Royal Name, that the soldiers shall not only be clothed and armed, according to the samples that will be sent herewith, & furnished with tents, and entertained in Our Pay, but that, after their arrival at the general place of rendezvous, they shall be provided for in every other respect as our British Troops; that they shall likewise have their just share and proportion of all plunder, or booty gained or taken from the enemy, according to their services, and if, by the Blessings of God upon Our Arms, any of the enemy's lands or settlements shall fall into our hands, our American Troops shall have a share thereof, as well as the rest of Our Forces, in preference to all other

persons that shall not have engaged in the present expedition. And we do further declare, that so soon as the present expedition shall be determined, whatever the success thereof may be, all the troops raised in Our American Colonies for this purpose, as well Officers, as soldiers shall have free leave to return to their respective provinces, and shall be transported thither at Our Expense, except those only, that shall choose, for their own advantage to settle or remain in such places or garrisons, as may fall into Our Hands, where they shall receive all due encouragement, and the soldiers, whether that remain there, or return home, after the said expedition, shall retain their arms & clothing." **Royal Instructions to Governor Ogle of Maryland, 2 April 1740**

Col. Blakeney, who is appointed Adjutant General in this expedition, will set out from here hence as soon as possible, with His Majesty's Letters and Authorities to you, and the other Governors of His Majesty's Colonies on the Continent of America....Col. Blakeney will carry with

> *him a considerable number of arms, some samples of clothing for the soldiers, and what money or credit shall be judged requisite for the performance of this service."* **Duke of Newcastle to Governor Thomas on the Expedition to Spain, West Indies, 5 January 1740**

The Crown expected the colonies to pay for support of the companies of the regiment prior to their arrival in Jamaica and to pay for their transportation. Major General Spotswood established the soldiers pay in "pistoles" which were convertible into local colonial currency at different rates, causing considerable confusion, especially since the colonies were also responsible for paying the soldiers until Colonel Blakeney assumed responsibility for the soldiers pay and reimbursed the colonies. In the 1740s, there still were, reportedly, Crown debts unpaid from a failed military expedition to take Quebec in 1711. Thus the colonies, specifically the Colonial Assemblies, had doubts that they would be reimbursed for the soldiers pay. The results were delayed payment to soldiers and ill will among the Royal Governors, Colonel Blakeney and the Colonial Assemblies.

> *"Recommend to you the utmost care and diligence in the execution of His Majesty's*

> *Orders; and particularly in procuring as great a number of men as you possibly can, to enlist ...you will see that Col Blakeny carries with him from hence only 3000 arms, you will not however limit the number of men to be raised within your government to, in proportion to that number of arms, since My Lord Cathcart will carry with him a quantity of spare arms and clothing, in case it should be practicable to raise a greater number than the three thousand for which Colonel Blakeney carries with him arms and money for their subsistence."* **Duke of Newcastle to the Deputy Governor of Maryland, 5 April 1740**

By 21 June, Colonel Blakeney had arrived with the Lieutenants coming from "North Britain" (Scotland). He brought with him 3,000 muskets, bayonets, cartridge boxes, 120 halberds and partisans, 60 drums and 500 uniform samples for colonial tailors to replicate. Camp colors and 100 tents were brought as well. Additional arms and uniforms for the Regiment were sent to Jamaica in anticipation of their arrival. Unlike other regiments in the British Establishment, the soldiers of Gooch's Regiment were not provided with swords. It is documentable that at least one company was sent to

Jamaica without being issued arms, uniforms and equipment. There is good reason to think there were many more. Colonel Blakeney was given approximately 800 Pounds as "equipage money" before departing England and there is no record of how that money was spent.

> *"Tents, bell-tents, drums, arms, ammunition and clothing for the American troops should meet them at the rendezvous in the West Indies on the supposition that they may be 4000 men, though no more need to be delivered at the rendezvous that the real numbers require."* **Calendar of State Papers Colonial, December 1739**

> *"New York, June 23. On Thursday night last arrived here Capt. Gill from London, but last from Portsmouth, who brings with him 16 Lieutenants for the expedition of the West Indies, 3000 fire arms, powder, tents, and other stores; there are 16 more Lieutenants on board of the Man of War from Portsmouth, with whom he parted on Sunday Morning last, the Man of War has Col. Blakeney, on board, and some Majors and Adjutants for the said expedition."*

New York American Weekly Mercury, 19 June 1740

"I have enclosed you commissions for three companies, the number of men at present expected from your government, and sent you the like number of Lieutenants and Sergeants with arms & powder and clothing for the Officers, so must desire you to form the men you have enlisted accordingly: Col. Blakeney has sent you bills to raise money, and I must entreat you to provide transports &c that I may be sure to have your men at the place of rendezvous within the Capes of Virginia before the middle of September." **Colonel Gooch to Governor Ogle of Maryland, 24 July 1740**

"August 8 Williamsburg, A vessel is arrived at Hampton, from New York, with several Officers, for the Companies raised here; as also with arms, ammunition, and clothing, for the soldiers." **New York Weekly Journal, 25 August 1740**

Colonial Governors were encouraged to raise additional companies over and above the 30 companies

Colonel Blakeney was coming to train, organize, and equip, which they did. The raising of additional Companies over the 30 planned created a need for additional commissions, which Colonel Blakeny was not empowered to provide. These additional commissions for company commanders, a total of six, four from Pennsylvania, one from Massachusetts, and one from New Jersey, took the form of warrants granted by Colonial Governors with the understanding that they would be recognized as commissions once the Regiment reached Jamaica. The newly raised men were to be equipped at Jamaica as well.

Prior to deployment to Jamaica, 15 newly raised companies disbanded; 6 from Massachusetts, 2 from Rhode Island, 5 from Connecticut and 2 from North Carolina. The men were unwilling to accept the assurances that their Governor granted warrants would be accepted as commissions for the officers and that the men would be properly clothed and equipped in Jamaica. The unwillingness of the men to accept assurances of valid commissions for American commanders and the resultant disbanded companies indicates a desire on the men's part to be led by those they knew and trusted. American soldiers insisting on being led by their own officers was nothing new and was a factor in the failed 1711 attempt on Quebec and continued to be an issue in the 1745 Siege of Louisburg and in the Great War for Empire in 1755.

"I was in hopes to have sent your Lordship ten companies from hence, by the proclamations I issued from his Grace the Duke of Newcastle's letters to me, wherein I promised the King's commissions to the officers & the King's arms to the men. But so it has happened that to the great disappointment of both no more commissions & arms than for four companies have been sent me. This so chagrined the men of six companies, whose officers had no commissions, nor they any arms, that they insisted the terms of their enlistment were not complied with, and refused to proceed. However, I have persuaded one company, under the command of Capt. John Winslow, to go forward without commissions & arms, which makes up five companies, instead of ten which would have gone had the commissions & arms came forward."
Governor of Massachusetts and New Hampshire Jonathan Belcher to Lord Cathcart, 23 September 1740

"Men would not list when they knew their officers could not be under the King's commissions, nor they have the King's

arms; & there was no persuading the men, neither there nor here, to wait for commissions & arms till they should join Lord Cathcart at the general rendezvous. I understand this important affair suffered for the same reason at Connecticut, where the Governor was obliged to dismiss five hundred, & the Governor of Rhode Island two hundred men. I have no doubt, may it please your Grace, whenever there may be a further occasion of this nature, upon the sending of commissions, arms, & clothing sufficient, that the men might depend to receive before their embarkation, I say, I have no doubt but six or eight thousand men might be raised in No. America; & this Province alone would easily furnish fifteen hundred men." **Governor of Massachusetts and New Hampshire Jonathan Belcher to the Duke of Newcastle, 29 October 1740**

The Regiment was mustered, for the most part, in two locations; Williamsburg and New York. Colonel Gooch commanded the Williamsburg detachment and Colonel Blakeney commanded the New York detachment. It is apparent that some training, at least sufficient for parades, took place in both locations prior to the deployment to Jamaica.

> *"New York Sept. 1. On Wednesday last arrived here.... on board of whom came the Hon. Col. Cope Lieutenant Colonel of the American Regiment commanded by Col Gooch. And in the evening of the same day the five Companies were mustered here before His Honor Our Governor, Col. Blakeney, and Col. Cope, after which they went through their exercise and made a handsome appearance."* **Boston Evening Post, 8 September 1740**

> *Williamsburg, Sept. 19. "On Monday last, the levies, that were encamped in this city, embarked on board the 4 transport ships that lay off Col. Burwell's, in James River; and on Tuesday and Wednesday they were paid their bounty money, three Pistoles to each man. Yesterday they set sail for Hampton, where they are to take in 50 or 60 men. There will be four complete companies, of 100 in each company, who by the care and diligence of their Officers, are disciplined in the new military exercise, and, it is hoped, not inferior to*

those of the other colonies...On Wednesday last, the Honorable William Gooch, Esq. Governor of this Colony, Quarter Master General, and Colonel of all the Marines raised in America, set out from this city for Hampton, in order to embark for the West Indies." **New York Weekly Journal, 20 October 1740**

"It was last Sunday before we arrived here having been by a most rapid current driven to Latitude 35-20. Col. Gooch to our great surprise sailed for Jamaica about a fortnight ago with the Maryland and Pennsylvania forces. The North Carolina Transports are not yet come neither shall we stay for them being ordered to sail to-morrow...As a spirit of dissatisfaction is natural with men in affairs of this kind I am afraid it will prevail with much reason to the dishonor and disgrace of our Country. The Expedition goes on with so much sloth and seems to be so ill concerted that I begin to be ambiguous whether it will go on or not." **Captain Provost, Hampton, Virginia, to his father, 26 October 1740**

"I can now assure your grace that we have raised 400 men in this province who are just going to put to sea. In those Northern Parts of the Colony adjoining to Virginia, we have got 100 men each, though some few deserted since they began to send them on board the transports at Cape Fear. I have good reason to believe we could have raised 200 more if it had been possible to negotiate the Bills of Exchange in this part of the Continent; but as that was impossible we were obliged to rest satisfied with four companies. I must in justice to the assembly of the Province inform Your Grace that they were very zealous and unanimous in promoting this service. They have raised a subsidy of 1200 pounds as it is reckoned hereby on which the men have subsisted ever since August, and all the Transports are victualed." **North Carolina Governor Gabriel Johnston to the Duke of Newcastle, October 1740**

> *"4 companies from Virginia, 8 from Pennsylvania, 3 from Maryland, all gone with Col Gooch; 5 from Boston, 2 from Rhode Island, 2 from Connecticut , 5 from New York, 3 from New Jersey, 4 from North Carolina – computed 3,500 men."*
> **Colonel William Blakeney at Hampton, Virginia to the Duke of Newcastle, 23 October 1740**

As previously noted, each colony was responsible for transporting the companies from their colonies to Jamaica, and feeding them until their arrival there. The degree to which they were supported while in transit varied from colony to colony.

> *"That the treasurer of this colony shall, and he is hereby empowered and required, out of the public monies now in his hands, and by borrowing so much as there shall be occasion for, in manner as hereafter is mentioned, to pay to the honorable William Gooch, Esq. his majesty's lieutenant-governor of this colony, the sum of five thousand pounds; to be by him applied towards providing victuals, transports, and other necessaries, for the soldiers, so as aforesaid raised, or to be*

raised in this colony, for his majesty's service, as herein before is mentioned." **Laws of Virginia, August 1740**

"We shall certainly be deficient in Provisions before we reach Jamaica having already expended about forty days and having about thirty yet to come. Some of our Assemblymen were pleased to tell me that the seventy days Provisions were to last ninety. I should be glad to have some of these Gentlemen on board at a pound of salt Beef for twenty-four hours which is the rate of their calculation according to man-of-war allowance as they say where they have rice with it and a change of fish, oil and oatmeal, all of which every vessel in the fleet has great plenty of save this one." **Captain Provost, Hampton, Virginia, to his father, 26 October 1740**

By December of 1740, Gooch's Regiment was assembled at Kingstown, Jamaica. Once there, Colonel Blakeney had problems paying for the soldier's rations. The contractors there refused to recognize Colonel Blakeney's authority to act on General Cathcart's behalf. Colonel Gooch made arrangements for funds to support the regiment with Governor Trelawney, much to Colonel Blakeney's chagrin. Unless they had been on sea voyages to the Caribbean, the tropical vegetation, sand, sea, not to mention heat, insects and diseases, were something the Americans had never experienced and, coming from temperate North America, could not have been prepared for. There is no record of training occurring while the Regiment waited for the arrival of Rear Admiral Ogle and the fleet carrying the other British Regiments participating in the upcoming expedition. The men, after a brief period of time spent on land in Kingstown, were placed back on ships and sent to a more remote part of Jamaica to await the arrival of the fleet from England. On 20 December 1740, Major General Thomas Wentworth succeeded Lord Cathcart as Land Force Commander, after Lord Cathcart died sailing to Jamaica.

> *"By letters from the Caribee Islands dated Dec. 24, it appears that Sir Chaloner Ogle arrived the 19th at Dominica and those transports which were separated from him by the storm, safely reached St. Christophers, the place of rendezvous, the 22d, which by mistake was in our last said to be Jamaica; whither Sir Chaloner Ogle proposed to proceed the 25th. Lord Cathcart died on the 20th, after 13 days*

illness, of the bloody flux, sincerely lamented by the whole fleet, and the command of the forces devolved upon Brigadier General Wentworth." **The Gentleman's Magazine, London, February 1741**

"The very day after their arrival at this island, the expedition sustained an irreparable loss, by the death of Charles, Lord Cathcart, a nobleman of distinguished valor, capacity, and experience in the military art, whose character was in all respects truly admirable, and whose fate was universally regretted, more especially as he was succeeded in command by Brigadier General Wentworth, an officer who had neither knowledge, weight, nor self-confidence sufficient to conduct an enterprise of such importance." **Tobias Smollett "An Account of the Expedition against Carthagena in the West Indies"**

On 9 January 1741, Vice Admiral Ogle and his fleet from England arrived at Jamaica. By then the conditions were taking their toll on Colonel Gooch's Regiment as seen in the following 11 January return for *"COL Gooch*

(Colonials) on transport ships at Jamaica: "deserted 45/ discharged 2/ sick 351 /dead since sailing 93/ total strength 3119.

4

Cartagena

In February 1741, Vice Admiral Vernon's fleet departed Jamaica with 186 ships, 29 being ships of the line, and 8,000 troops, including approximately 3,000 men of Gooch's Regiment. The fleet first sailed to Hispaniola, and then made for Cartagena. They stopped on the way for "water and wood" from 17 to 25 February. The following gives a sense of how the soldiers of Gooch's Regiment were perceived and the type of work they could expect to perform in the high heat and humidity of the tropics for the rest of the Cartagena expedition.

> *"February the 16th, it was resolved, in general council of war, that the fleet should, with all expedition, wood and water in Rios, Tiberoon, and Donna Maria bays, and thence proceed to Carthagena. During the seven days which were thus employed, detachments from the American regiment and the negroes were daily sent*

ashore to cut fascines and pickets."
Tobias Smollett "An Account of the Expedition against Carthagena in the West Indies"

By 4 March, Vice Admiral Vernon's entire fleet arrived off the coast of Cartagena. The returns indicate that for the period February 1741 to July 1742 Captain Washington and 144 men from Gooch's Regiment were on board the HMS York. Information of this sort will be included in the narrative and may seem incidental at this point, but will provide the almost sole evidence of Gooch's Regiment's participation in future operations.

"Col. Gooch's Regiment of Foot's" returns at "Carthagena Harbor," 23 February 1741

	On Transports	*On Men of War*	*Sick at Jamaica*	*Totals*	*Short*
1st Bn	*62*	*618*	*67*	*775*	*125*
2nd Bn	*105*	*638*	*40*	*777*	*121*
3rd Bn	*184*	*524*	*37*	*745*	*155*
4th Bn	*124*	*669*	*21*	*815*	*85*

On 9 March 1741, an initial landing was made on the west side of Tierra Bomba, north of Boca Chica. It would not be until late March that a local source of fresh water was found, until then all water and food had to be

transported to the soldiers in an environment where daytime temperatures would have been in the 80-90 degree Fahrenheit range and humidity would have been equally high.

> *"People on board the fleet, who had been hitherto restricted to a very scanty allowance of this element, namely, a purser's quart (about three half pints) per diem to every individual; in a climate where there is such a continual expense of animal fluid, that as many gallons might have been necessary to repair the waste of four and twenty hours, in a hard working man, sweating under the sun, which was vertical, and fed with putrid beef, rusty pork and bread swarming with maggots."*
> **27 March 1741, On the discovery of a spring, Tobias Smollett "An Account of the Expedition against Carthagena in the West Indies"**

On 10 March a second contingent was landed.

> *"Next morning the grenadiers were formed on the beach, to cover the landing of the rest of the army, which was effected without opposition; though the troops*

were obliged to lie that night upon their arms." **The events of 10 March 1741, Tobias Smollett "An Account of the Expedition against Carthagena in the West Indies"**

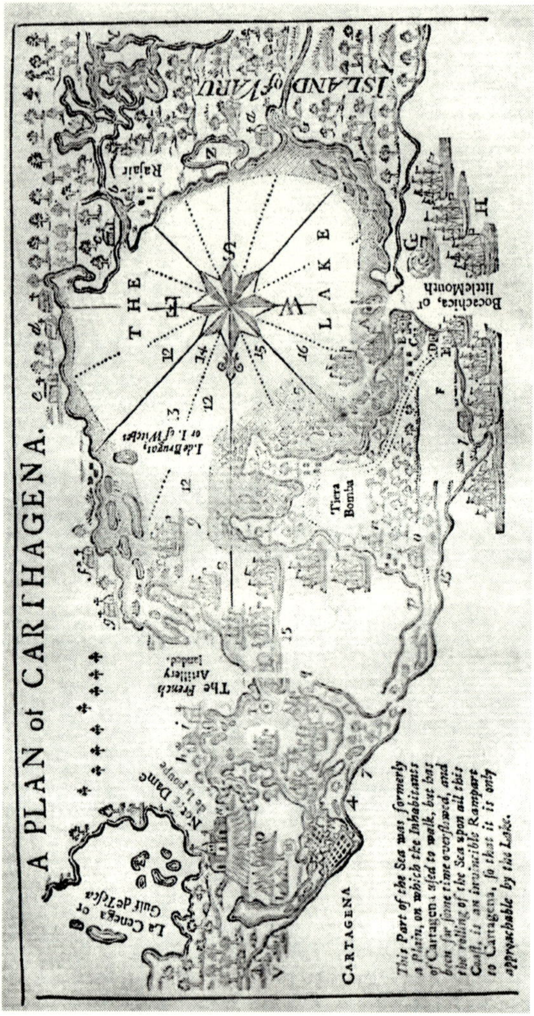

"A Plan of Cartagena," The Gentleman's Magazine, London, April 1741. ***Bocca Chica, the entrance to the inner harbor is located due west. Barradera is to the south of the narrows and Casle San Luis is to the north. Castle San Lazario is north east of Cartagena.***

The troops remained on the beach for a period of three days without improving their position, for no apparent reason in the literature. The tropical beach upon which they landed was within sight of the jungle, but offered no protection from the sun for the wool uniformed soldiers. Intermittent breezes would have kept the disease bearing mosquitoes away but when the breezes stopped, the mosquitoes swarmed. This period of exposure had a serious adverse effect on their health.

> *"During the first three days the troops were ashore, they were employed in no one thing, no not so much as to clear the ground for their encampment, but kept under arms night and day (where, by the heat of the sun, on a white burning sand, they were scorched to death, and by the inclemency of the dews in the night, they got colds, so that many of them fell sick) whereas had they been instantly employed to have encamped and opened ground in the woods for that purpose, they would have been shaded by the trees, freed from the burning heat of the sand, and many of them preserved from the enemy's shot that missed our battery."* **Events of 10-12 March, 1741 according to Charles**

Knowles, "An Account of the Expedition to Carthagena"

On 15 March a request for more troops from Major General Wentworth to Vice Admiral Vernon was refused, beginning a pattern of conflict over the men of Gooch's Regiment who were being called on to replace sick and dead sailors as well as to serve in the manner they had volunteered for, as soldiers.

> *"The principal Engineer not arriving till the 15th, no spot was pitched upon for raising a battery against the enemy, so that the clearing a few bushes away down by the water side, for to pitch their tents was all the material work the army did for near a week."* **Charles Knowles, "An Account of the Expedition to Carthagena"**

> *"The engineer represented to the council for war, that the work could not be carried out in a proper manner, without a reinforcement of one thousand six hundred men added to the troops already landed; and as there was a number of soldiers still on board of the fleet, the general demanded this supply; which the admiral*

refused on the supposition that such a reinforcement could not be necessary."

Events of 15 March 1741, Tobias Smollett "An Account of the Expedition against Carthagena in the West Indies"

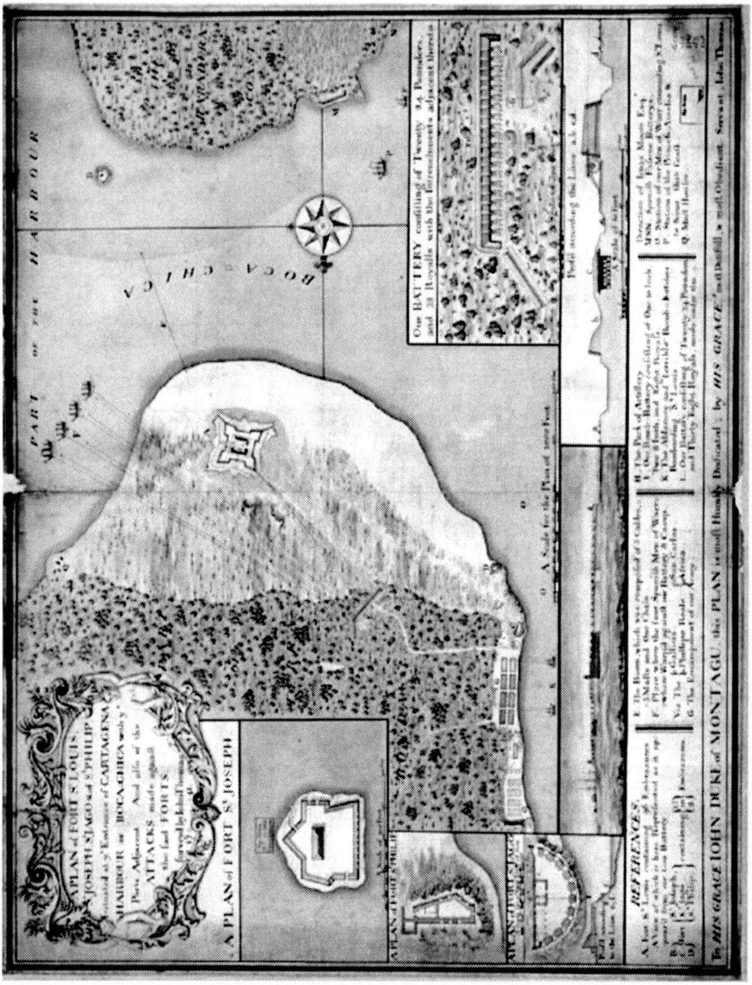

Map of Operations at Cartagena, 9-25 March 1741 by John Thomas for the Duke of Montagu. Castle San Luis (A) is to the north east and the British battery (I) and camp (G) is to the north west. The Barradera battery (M) is south of Boca Chica.

It is clear that the soldiers of Gooch's Regiment who were permitted to serve with the land force, were used for the most part as laborers, rather than soldiers. From 16 to 22 March, a 200 man element of Gooch's Regiment, along with the armed Jamaican troops constructed trenches and a battery in preparation for a siege of Castle San Luis De Boca Chica. There is nothing in the record to indicate any of the other regiments participated in the preparation of earthworks in this tropical jungle.

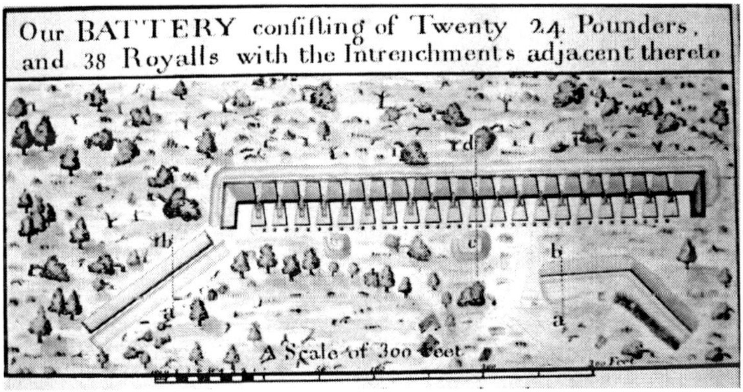

Detail from the John Thomas Map. An overhead view of the Battery built by Gooch's Regiment 16-22 March 1741. Note scale at bottom of image.

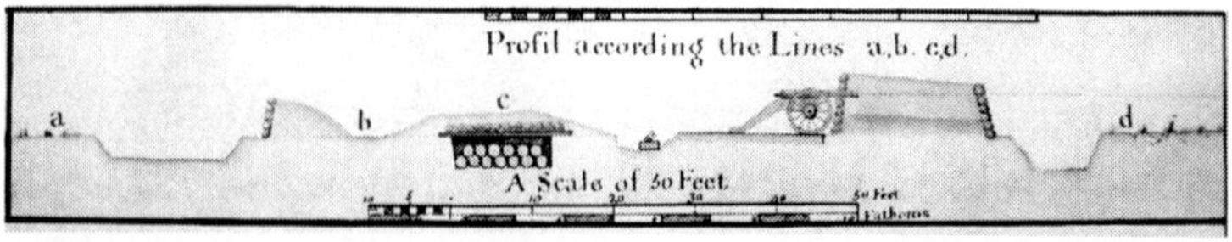

Detail from the John Thomas Map. A cross section view of the Battery built by Gooch's Regiment 16-22 March 1741. Note scale at bottom of image.

Captain Lawrence Washington, the elder half brother of George Washington, enters the account with his participation, along with soldiers of Gooch's Regiment, in an amphibious raid to destroy a fascine battery on Barradera, 18 March 1741. The raid was a great success and no doubt improved the morale of the soldiers who had been under fire from the ruined Spanish Battery.

> *"March the 17th, when the parapet of the battery was raised almost to the embrasures, the land officers, in a council of war, agreed to solicit the assistance of the admiral in destroying a fascine battery on the other side of the harbor, called the Barradera, from which they had received considerable damage and interruption; and in compliance with this request, a detachment of three hundred sailors, supported by a body of soldiers that still remained on board of the fleet, were*

conveyed thither at night in boats, under the command of Captains Boscawen, Watson, Coats, Washington, Mr. Murray, and Lieutenant Forrest, who attacked the battery with great valor, repulsed the enemy, and spiked up the cannon."
Tobias Smollett "An Account of the Expedition against Carthagena in the West Indies"

"And for carrying it on with more secrecy, that it ought to be done with the barges and pinnacles of the fleet only; and that twenty five men from each ship of the line might be embarked for the said service, besides the boats crew to be fifteen seamen and ten soldiers, the seamen armed with swords, pistols, cutlasses and hand-grenadoes for rushing in upon them; and the soldier with their firelocks and bayonets to support them and secure their retreat." **Resolution of a Council of War held on board His Majesty's Ship, the Jersey, March 17th, 1741**

"They went away from their ships about midnight, and rowed pretty far to leeward, to avoid being seen, or the noise of their

oars heard, and proposed landing in a small sandy bay, behind the Barradera Battery, into which was a narrow channel (both unknown to every person there) which, so soon as some of the boats had got into the channel, began to fire on them; but the brave tars landed, and rushed in at the embrasures, and took possession of the battery, before the enemy could fire a second time. This firing alarmed the Barderra Battery, and the enemy turned three pieces of cannon on the platform, which they fired with grapeshot, so soon as the seamen advanced; but not withstanding that, and the difficulties and badness of the road (which was through a morass, and where but one man could walk abreast, and full of stumps of mangrove each a foot or more high, the seamen attacked it; and, after a smart though short resistance, carried it, took nine prisoners, spiked up fifteen guns (from eight to twenty pounders) burned the carriages, platforms, guardhouses and magazine." **Charles Knowles, "An Account of the Expedition to Carthagena"**

There was a second amphibious raid on 24 March, but there is no record of participation by Gooch's Regiment.

> *"In the evening, another detachment of sailors and soldiers, commanded by Captain Waters, was transported in boats to the Barradera shore, and burned the battery without opposition, together with a sloop which lay on the other side of a neck of land."* **Events of 24 March 1741, Tobias Smollett "An Account of the Expedition against Carthagena in the West Indies"**

On 25 March 1741, Castle San Luis De Boca Chica was stormed with support from the 20 twenty-four pounders of the attached artillery commanded by Lieutenant Colonel Jonas Watson. No longer defended by Spanish forts, the boom across Cartagena harbor was broken allowing access to the inner harbor. Once that occurred, an additional contingent of 1600 men was landed, 200 of which were from Gooch's Regiment. Based on the following account, and the fact that Colonel Gooch was wounded in this action, it is probable that Gooch's Regiment participated in the storming of Castle San Luis De Boca Chica.

"Another Letter from before Cathagena, dated April 3, 1741. Boca Chica was taken the 25th of March, when there was but a small breach made, being stormed by one sergeant and 12 soldiers, the forlorn hope of the army. Officers killed My Lord Averbury, two Colonels, one Captain, one Lieutenant, with between two and three hundred common soldiers...Our northern forces have behaved gallantly; Admiral Vernon calls them his brave Americans. Henry Hodge." **Boston Evening Post, 1 June 1741**

"But before they reached the foot of the walls, they heard the Spanish drums beating to arms; the top of the breach was manned, and the enemy's ships, and Fort St Joseph, began to fire with grape-shot upon the assailants, though with very little effect; but the commandant, Don Blas, being on board one of the ships, the garrison were struck with a panic, and fled with precipitation out of the gate as soon as the grenadiers began to mount the breach." **Events of 25 March 1741, Tobias Smollett "An Account of the**

Expedition against Carthagena in the West Indies"

"People on board the fleet, who had been hitherto restricted to a very scanty allowance of this element, namely, a purser's quart (about three half pints) per diem to every individual; in a climate where there is such a continual expense of animal fluid, that as many gallons might have been necessary to repair the waste of four and twenty hours, in a hard working man, sweating under the sun, which was vertical, and fed with putrid beef, rusty pork and bread swarming with maggots."
27 March 1741, On the discovery of a spring, Tobias Smollett "An Account of the Expedition against Carthagena in the West Indies"

The following show the ongoing struggle between Vice Admiral Vernon and Major General Wentworth over the men of Gooch's Regiment. Vice Admiral Vernon continually refused to land the men of the Regiment serving on board the Men of War who were serving as replacements for sailors.

"It was resolved he should be supplied from the men of war, with all such of the American forces, as he should judge proper to be trusted on shore." **Resolution of a Council of War held on board The Princess Carolina, Carthagena Harbor, March 30th, 1741**

"A detachment of fifteen hundred men, under proper officers, from the American Regiments, commanded by Colonel Gooch, be held in readiness to join the troops now on board the transports, on their landing." **A Representation from the Commander in Chief and the Major General of the Land Forces, to Vice Admiral Vernon, April 2, 1741**

"The service will require, that the remainder of the troops be landed as soon as shall be possible, not only those that were ashore at Terra Bomba, but the fifteen hundred Americans." **Copy of a letter from Major General Wentworth to Vice Admiral Vernon, Dorsetshire-Transport, April 4, 1741**

On 5 April 1741, landings were made within striking distance of Castle San Lazario, a fortification that controlled the inner harbor of Cartagena. A contingent of Gooch's Regiment covered the flank of the main attacking column deployed as skirmishers in the jungle. It is interesting to note the use of Americans in a manner similar to the close-in woods fighting characteristic of combat in North America.

> *"April the 5th, the first division of the soldiers, amounting to one thousand four hundred men, commanded by Brigadier Blakeney, rendezvoused along side of the Weymouth; at five in the morning, Colonel Grant, at the head of the grenadiers, landed without opposition, and being followed by the rest formed the whole in order of battle, to march against the enemy. These being joined by two hundred Americans as pioneers, the negroes and a party of matrosses, with eight patteroes, began to move forwards into the woods, and having reached the end of the defile, with the loss of one man killed by the fire of a straggling party, they perceived the Spaniards, to the number of seven hundred, drawn up on the strand, in such a manner as to cover the road*

leading into the city. There they stood with a good countenance, seemingly determined to dispute the ground, from which the general ordered the grenadiers to dislodge them; but, as they were obliged to march through a narrow pass, bounded on the left by the lagoon and on the right by a thick copse, a party of American soldiers were detached into this last, in order to fall upon the rear of any small parties which might have been posted among the bushes, to flank the English as they advanced." **Tobias Smollett "An Account of the Expedition against Carthagena in the West Indies"**

The events of 6 April still show Gooch's Regiment being used as a source of labor as additional forces were landed in preparation for operations against Castle San Lazario, called Castillo San Felipe De Barajas by the Spanish. Vice Admiral Vernon was still resisting Major General Wentworth's efforts to land the soldiers of Gooch's Regiment serving on shipboard.

"That very evening, however, five pieces of cannon with powder and ball, were put on shore, and a number of Americans and negroes being landed, with working tools,

began to clear the ground for an encampment; though in the mean time the Europeans suffered severely from excessive heat; so that the work was greatly retarded." **Events of 6 April 1741,Tobias Smollett "An Account of the Expedition against Carthagena in the West Indies"**

"Part of the Americans are joined us; you will be pleased to order that those from Mr. Lestock's Squadron be landed by the first opportunity." **Copy of a letter from Major General Wentworth to Vice Admiral Vernon, La Quinta, April 6, 1741**

"From the first sight of the American troops they were despised, and as many of them were Irish (suspected Papists) were never employed till now; but as sickness increased amongst the others (and hourly attacks expected from the enemy) it was thought expedient to have them ashore; and though it is most certain, there was scarce one but knew what opinion had been conceived of them, nay indeed told them,." **7 April, 1741, Charles Knowles,**

"An Account of the Expedition to Carthagena"

The record of council of war resolutions shows that the Land Forces Council of War on 7 April called for a siege of Castle San Lazario "by form", involving trenches and artillery batteries creating a breach for an infantry assault. The record is silent between the period of 7 April and 9 April when an assault was ordered without benefit of a breach in the walls. There are at least two secondary sources that describe Colonels Blakeney and Wolfe as objecting to the attack of 9 April.

> *"It appears to the said council of war that the walls of the Castle of St Lazar cannot be less that twenty-five feet in height, which is equal to the length of their longest ladders, that there is, besides, a ditch with water. and very muddy at the bottom, supposed to be about fifteen feet in breadth and about six feet in depth: we therefore, for these reasons, deem it impracticable to scale the walls: and do resolve, with all possible dispatch, to raise a battery, in order to make a breach."* **At the Headquarters at La Quinta, April 7th 1741**

The assault on 9 April was made through a narrow pass allowing only for a four man front. The Spanish did not flee as they had at Castle San Luis De Boca Chica and perhaps the British expected them to, precluding a need to breach the walls before assaulting. The conduct of the soldiers from Gooch's Regiment that participated in the assault is the source of some controversy.

> *"Ye enemy kept an incessant fire on us while we advanced & at 4 we got under ye walls &began our fire which we continued till 8 of ye clock when we found it impracticable to continue any longer& was obliged to retreat with great precipitation while our enemy kept firing on us, we had killed or wounded near 700 men but ye greatest part kill'd dead on ye spot we was so near ye walls that we could hear our enemies say in English words God Damn you why don't you lower ye muzzles of ye guns lower ye muzzles & fire away I am sorry that some of our Subjects Eyes are not as yet open."* **Description of the assault on St Lazar, Journal of Captain John Lloyd, Gooch's Regiment**

"The 9th of April, about 4 o'clock in the morning, a body of men, consisting of 500 Grenadiers, supported by 1000 European Marines and Jamaican soldiers, marched towards the enemies lines before Fort Lazar, and were followed at a considerable distance by a large body of Americans, laden with wool-packs, scaling ladders, and hand grenades….….The Americans, finding they were knocked down without any arms to defend themselves, threw down their ladders, &c. and retired to their camp. Three only were brought up to the trenches, upon which about ten of our Grenadiers and a Sergeant mounted upon the walls of the Fort; but being unsupported, were immediately cut to pieces, except the Sergeant who saved himself by jumping down, again. Fort Lazar is a strong Castle of about 30 guns, the walls built of hewn stone 40 feet high, situated on the summit of a hill, and environed with trenches and fascine batteries." **Boston Evening Post, 9 November 1741**

Assigned the mission of bringing forward the ladders, already identified at the 7 April Council of War as too short, soldiers dropped their ladders in order to arm themselves and return fire because those assigned to provide covering fire to the ladder carriers failed to do so. Some sources credit Captain Washington as having participated in the assault. Captain Addison participated in the attack and Lieutenant Marshall of Captain Clark's Company was wounded in the assault.

> *"Nor could the scaling-ladders, wool-packs, or hand grenades be of any service in this emergency, for the Americans who carried them in the rear, seeing the troops falling by whole platoons, refused to advance with their burdens; but though they would not advance as pioneers, many of them took up the firelocks which they found on the field, and, mixing among the troops, behaved very bravely."* **Events of 9 April, 1741, Tobias Smollett "An Account of the Expedition against Carthagena in the West Indies"**

The failed assault on Castle San Lazario was the last ground operation against Cartagena. The onset of the nearly constant rains if the rainy season led to an increase in sickness among the soldiers and sailors of the expedition leaving no choice but to depart Cathagena.

> *"It was therefore found necessary to effect a retreat which was secured by means of those five hundred who brought up the rear; but not before the loss of the English amounted to two hundred killed and twice that number wounded, of which the vast majority did not recover."* **Events of 10 April 1741, Tobias Smollett "An Account of the Expedition against Carthagena in the West Indies"**

> *"The sickness still increasing among the troops, and the admiral declining to land a reinforcement for supplying the loss by which the army was diminished; it was resolved, in a council of war, to desire the admiral to give directions for re-embarking the cannon, since his silence, with regard to the demanded reinforcement, seemed to imply denial."* **Tobias Smollett "An Account of the Expedition against Carthagena in the West Indies"**

"At the time this council of war was assembled, the troops were by sickness and loss reduced to three thousand five hundred and sixty nine men; and many of these feeble and worn out with fatigue, as scarcely to be fit for the ordinary duty. In the number above is included, one thousand and one hundred and forty American soldiers, six hundred of whom were only employed as working parties."
Events of 14 April 1741, Robert Beatson, "Naval and Military Memoirs of Great Britain, from 1727 to 1783"

"Honorable Colonel Gooch's Regiment of Foot's" returns at "Carthagena Harbor," 23 April 1741

	On Transports	*On Men of War*	*Sick at Jamaica*	*Totals*
1st Bn	*98*	*571*	*65*	*734*
2nd Bn	*125*	*576*	*33*	*734*
3rd Bn	*270*	*392*	*37*	*669*
4th Bn	*116*	*664*	*21*	*801*

	Short	*Sick*	*Dead*	*KIA*	*WIA*
1st Bn	*166*	*20*	*33*	*10*	*29*
2nd Bn	*166*	*98*	*6*	*17*	*17*
3rd Bn	*201*	*129*	*14*	*9*	
4th Bn	*99*	*197*	*13*	*3*	*21*

> *"The general in person brought up the rear; and perceiving that five tents belonging to the Americans were left standing; and some tools lying on the ground, he ordered them to be fetched by a sergeant's guard reinforced by some sailors by Lieutenant Forrest; so that there was nothing left as a trophy to the enemy, who did not think fit to molest them in retreat."* **Tobias Smollett "An Account of the Expedition against Carthagena in the West Indies"**

On 12 May, 1741 the British fleet departed Cartagena, arriving at Jamaica on 19 May 1741. From the following, we see there was little concern about where the surviving soldiers of the Cartagena Expedition were bivouacked.

> *"The surviving troops were landed and encamped at Greenwich, a mile or two to*

the west of Kingston. Here the Rio Cobre and other streams discharge themselves into the shallow, muddy waters of Hunt's Bay, and all around are swamps. It was May; the usual rains were very heavy, the ground at Greenwich was a bog; and here, as if British soldiers were worthless chattels, they were left to gaze upon the glorious healthy mountains but a mornings march to their rear, to listen to the ripple of the waves on the muddy shore, until sight and hearing failed, and death released the sufferers." (Gardner, "A History of Jamaica," p. 124)

"We in short, destroyed eight forts, six men of war, six galleons, and some merchant ships; what number of men they lost we know not; the enemy killed of ours about six hundred and some wounded, and the climate killed us in greater numbers. Vast changes we have in each regiment; some are so weak as to be reduced to a third of their men, a great quantity of officers amongst the rest are dead. Col. Gooch received a wound but is very well recovered. What we are next to do is a secret, some talk of Havana, other

Panama, some Vera Cruz but perhaps others more probably Santiago Cuba. We are all tired of the heat and wish for a cold season to refresh our blood. I mentioned the exchange of this island before but they have now raised the price of everything.... I have sailed on Admiral Vernon's ship since we left Hispaniola vastly to my satisfaction....Our Regiment has not seen the treatment we expected, but I am resolved to persevere in the undertaking. War is horrid in fact, but much more so in our imagination; we there learned to live on ordinary diet, to watch much, to disregard the noise or shot of cannon...Most people imagine our Regiment placed on the very best footing and that we shall be continued in North America, but I hope not at the expense of the Colonies. Some talk of recruiting in America. If so I shall apply for one and there do the best I can to inform you of the affair which however is so intricate and contradictory that I do not care to assert any particulars, the Sea and Land forces relate it differently. If the sailors can claim much honor I partake of it as Capt of the soldiers who acted as Marines; but we meet no opposition though really some of the attempts were bold or rather rash. Most of our American Transports are discharged wherefore imagine it no sign of our speedy return." **Captain Lawrence Washington at Jamaica to Augustine Washington, 30 May 1741**

"It is a melancholy truth, which however, ought to be told, that a low, ridiculous, and pernicious jealousy subsisted between the land and sea officers during this whole expedition; and that the chiefs of those were so weak or wicked as to take all opportunities of thwarting and manifesting their contempt for each other, at a time when the lives of so many brave fellow-subjects were at stake, and when the interest and honor of their country required the utmost zeal and unanimity." **Tobias Smollett "An Account of the Expedition against Carthagena in the West Indies"**

"Saturday last, the 20th instant, the brave Capt. Warren arrived here from Jamaica, by whom we are informed that Admiral Vernon, with his fleet, and General

> *Wentworth, with his land forces, have left off the siege of Carthagena, on account of great sickness and mortality among their men, and are arrived at Jamaica. While they lay in their trenches before Carthagena it rained ten days together, during which time there died (it's said) above one thousand of their men. And they have lost in all, about 8000 men .....The officers and soldiers from North America behaved themselves with much bravery, but many of them are dead."* **25 June 1741, Boston Weekly Newsletter**

During the period of June 1741 to December 1742 the HMS Hamilton Court sailed with 23 men of Gooch's Regiment. In June of 1741 Colonel Gooch, departed Jamaica for Virginia recovering from wounds sustained in both legs during the siege of Cartagena. Lieutenant Colonel Cope, the Regimental Executive Officer, assumed command. The four battalions in Gooch's Regiment were also reorganized with three companies each.

> *"Williamsburg, Virginia, July 31, On Monday last, the Hon. Col. Gooch our Governor, arrived at York, in the ship Buchannan...we were agreeably surprised*

with His Honor's return to this City, in good health, though somewhat lame of the wounds he received by a cannon ball, in his legs before Boca Chica." **Boston Evening Post, 24 August 1741**

5

Cuba

"Admiral Vernon, having refitted, and the Land Forces under General Wentworth, being recovered and re-embarked, set sail with the fleet under its command, and 41 transports, on the first of July, and on the 18th anchored in Waltheham Bay at the S.E. of the Island of Cuba to which he gave the name of Cumberland Harbor."
Samuel Boyse, "An Historical Review of the Transactions of Europe"

In July of 1741, Vice Admiral Vernon and Major General Wentworth left Jamaica with 2158 men, including Gooch's Regiment. Their plan was to establish a base of operations on the Island of Cuba in the vicinity of the Port of Santiago (modern day Santiago De Cuba) which was the largest settlement on the island at the time. The conditions in Cuba would not be too different from those of Cartagena with similar high heat, humidity and life threatening disease.

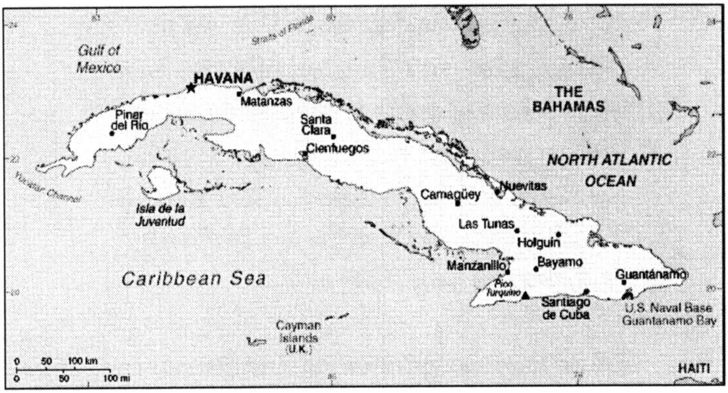

In 1742 Santiago De Cuba was called Santiago and Cumberland Harbor was located at the current U.S. Naval Station

The only combat operation described while the British forces were on Cuba is the attack on the village of Catalina on the Santiago Road, which included 2 companies from Gooch's Regiment. One of the companies from Gooch's Regiment was commanded by Captain Washington.

> *"We having on the 25th sent out a detachment of a hundred of the American troops and a hundred blacks, I ordered Lieutenant Sturton from my ship with ten men from my brig tender, to take the guide under his care, and to accompany them. And on the 26th at night I received from General Wentworth an account of their*

success; the enemy's first advanced guard flying before them without so much as firing a shot with them, and leaving them eleven horses, some of their ammunition and a good deal of jerked beef." **Vice Admiral Vernon on the HMS Boyne in Cumberland Harbor to the Duke of Norfolk, 29 July 1741**

"You are hereby required and directed to take with you a detachment of thirty men of Capt. Pepper's Company, now on board with you, with the said Captain, and a further detachment of twenty American forces, under one of their officers, from the Montague and Ludlow Castle, to be landed, for marching under your orders, by day break tomorrow morning, at the bluffs to the westward of the river: from whence, together with 20 seamen (with hand bills, for clearing underwood where necessary, and armed with sword and pistols only, as you will have sufficient number of musketeers from the detachments of American troops) you are to march very leisurely, for opening your communication with the said encampment on the river." **Vice Admiral Vernon on**

the HMS Boyne in Cumberland Harbor to Captain Rentone, of the HMS Rippon, 6 August 1741

"We have suffered and still suffer, but that our stay in this place has already brought us into great difficulties and are of the opinion that unless a speedy resolution be taken to extricate us it will be out of our power to retreat from hence and rejoin our army; the river which separates us from Major Dunston's party grows sensibly deeper every day and Captain North, Col. Lewis and Mr. Innes who came from camp yesterday assure us that the rivulet which hardly covered the shoes when we first passed it is now mid thigh deep." **Letter to Colonel Whitefoord from the Officers under his command Cumberland Harbor, 11 August 1741**

"Your first advanced party was made up of, to a very few men, of the Americans I sent up in my flotilla a hundred blacks; and as to the company from this ship, I know they were mostly composed of Irishmen, and some of them convicts; Yet, by the report of Capt. Washington, their

Captain, they all went on service with great cheerfulness, and are all returned according to your orders, without a man deserting; and they were concerned to be recalled and expressed themselves desirous of going again." **Vice Admiral Vernon on the HMS Boyne in Cumberland Harbor to Major General Wentworth, 19 August, 1741**

Major General Wentworth's men were withdrawn back to the harbor and the Spanish took no action, knowing that the climate over time would solve the problem created by the British land force at Cumberland Harbor. According to Spanish sources, they were prepared to defend the road with hit and run tactics should the British advance down the road.

	Serj.	*Corp.*	*Drums*	*Sold.*
1st Battal.	*16*	*11*	*3*	*129*
2nd Battal.	*10*	*7*	*3*	*90*
3rd Battal.	*10*	*7*	*3*	*79*
4th Battal.	*6*	*7*	*1*	*107*

Gooch's Regiment from *"Strength of our Army In Cuba under General Wentworth Oct. 5, 1741"* ***Gentleman's Magazine, February 1742***

"Capt. Washington, who was here from Col Cope, to have an account of the particular disposal of Americans on board our ships, had from me a particular account of them, as he desired, and you know, sir, their making a confusion, by sending them on board by detachments from each company, was their own handiwork." **Vice Admiral Vernon on the HMS Boyne in Cumberland Harbor to Major General Wentworth , 9 October, 1741**

"It appearing by a return (transmitted by Major General Wentworth, and laid before the Lord Justices) of the State of the American Regiment, under his Command, that several private soldiers belonging to the said Regiment have been impressed by mistake on board His Majesty's Ship of War, and detained, not with standing application had been made for their being restored to their proper officers; and that their excellencies considering, that, should these men be treated contrary to the promises made them at their enlisting as soldiers, it might discourage His Majesty's Subjects, in America, from entering into

> *His Service, they have commanded me to signify to you their directions, that you should order the soldiers above mentioned, and all others of the American Regiment, that may have been impressed on board any of His Majesty's ships, to be immediately discharged and sent back to their Regiment."* **Order to Vice Admiral Vernon 16 October 1741** (Leach's "Roots of Conflict" p.58)

Vice Admiral Vernon was continuing his practice of using the soldiers of Gooch's Regiment as replacement sailors. In his letter to the Duke of Newcastle he repeats the unsubstantiated assertion found in Newcastle's letter to the Governor of Pennsylvania of January of 1740 that the Americans wanted to settle in the Caribbean.

> *"I think my inclinations have been entirely conformable, to what, I believe, was the principle motive of all the American Officers engaging in the service, the hopes of being settled in the West Indies, and in Cuba preferably to all places. But I have heard some of them mention many complaints they had to make of their treatment in the army; and of the discontents expressed by some of the*

> *Europeans (as the Americans called them) that they were to expose their lives for procuring settlements for the Americans."* **Vice Admiral Vernon on the HMS Boyne in Cumberland Harbor to The Duke of Newcastle, 3 November 1741**

In November 1741, the base at "Cumberland Harbor" was abandoned with only 300 soldiers fit for service out of the original 2158 that landed on Cuba.

> *"For the better securing the re-embarkation of the forces from the camp; you are hereby required and directed to receive on board his Majesty's Bomb Vessel the Alderney from the Kent, a Lieutenant, two Sergeants, two Corporals, a drummer, and twenty four private men, belonging to Colonel Gooch's Regiment, and bear them for victuals only while they continue on board."* **Vice Admiral Vernon on the HMS Boyne in Cumberland Harbor to Capt. Forrest, 10 November 1741**

> *"All the cannon, baggage, provisions, and sick men, being embarked, on Monday the*

16th in the morning we discerned the huts of the camp to be on fire, Mr. Wentworth having that morning marched down with his remaining well men, and embarked himself." **Vice Admiral Vernon on the HMS Boyne in Cumberland Harbor to The Duke of Newcastle, 26 November 1741**

"A Regiment of Foot raised in America, 3763 men", Establishment of the Land Forces for the Year 1741," **The Gentleman' Magazine, London, November 1741**

"Colonel Gooch's Regiment of Foot" returns at "Kingston in Jamaica," 7 December 1741

	Fit	*Sick*	*On Men of War*	*Totals*
1st Bn	*42*	*204*	*267*	*480*
2nd Bn	*36*	*193*	*253*	*445*
3rd Bn	*56*	*89*	*251*	*381*
4th Bn	*62*	*183*	*289*	*509*

	Short	*Dead*	*Deserted*
1st Bn	*421*	*69*	*2*
2nd Bn	*455*	*34*	
3rd Bn	*519*	*28*	
4th Bn	*391*	*88*	*1*

"I was exceedingly surprised upon reading Capt. Trevor's letter, that you had prevailed on him to give orders, for the sick soldiers of the American Regiment being received into the hospital for our seamen at Port Royal: an order, I conceive, that neither he nor I can justify, it bringing such a charge on the Navy, and which he may be liable to have charged upon his wages.

Whereas, Sir, you well know, you have an established hospital for your Army, and much more numerous Officers and Surgeons appointed for the care of it than we have, though we have a greater number of men to be taken care of; and you likewise have the strictest orders, for the American Regiment to be regarded in every respect as the other forces your Army is composed of, and have the large town of Kingston to provide quarters for them all........I must beg the favor, Sir, you will give immediate orders for the sick American soldiers, lent to us to man our ships, being received into the Army Hospital whenever they have occasion for it." **Vice Admiral Vernon on the HMS**

Boyne in Cumberland Harbor to Major General Wentworth, 5 January, 1742

In February 1742 Major General Wentworth's land force was reinforced with a battalion from the Royal Scots, the 6th Foot and the 27th Foot, for a total of approximately 3,000 men. Attrition of the new arrivals due to disease began almost immediately. This is demonstrated by the return for the period 20 January to 23 February showing the death of 109 men from the "old" European Regiments, 165 men from the "new" European regiments and 99 Americans.

6

Recruiting for Replacements

In August 1741, Major General Wentworth sent Captain Winslow and Captain Hopkins to the Mid Atlantic and New England Colonies to recruit for the entire expeditionary land force.

> *"His Majesty having directed me to use my best endeavors to recruit his forces under my Command , in his colonies in North America, I have appointed Captain Hopkins to repair for that end to your province, not doubting of your giving him all the assistance in your power, to levy soldiers not only for completing Col. Gooch's Regiment, but if practicable to raise a greater number, either to fill up vacancies in the two old corps, and in the Marines, or to form another battalion, as it shall be found best for his Majesty's service...All such as shall enlist themselves on this occasion, will be*

> *entitled to the advantages offered by his Majesty, in his Instructions on the first raising Col. Gooch's Regiment."* **Major General Wentworth at Cuba, to Governor Ward (of Rhode Island), 12 August 1741**

> *"You are as soon you land to repair to such provinces as you think most likely to afford you recruits, of which you are to raise as many as you shall be able with all expedition...You are only to enlist able bodies men, not exceeding thirty five years of age, nor are you to enlist any native of Ireland, except you have undoubted proof of his being a Protestant, and of good character."* **Major General Wentworth's Orders to Captains Hopkins and Winslow, 12 August 1741**

There was no effort to recruit any more men from Maryland, Virginia or North Carolina.

> *"None to be expected from Virginia, Maryland or North Carolina, who are not Irish, Papists or English convicts of whom we have but too many in these four battalions."* **Colonel William Blakeney at Jamaica to The Duke of Newcastle,**

on recruitment in the North American colonies for Gooch's Regiment, 2 June 1741

As a result of these recruiting efforts Pennsylvania sent 92 men, Massachusetts 50, New York 20, and Rhode Island 53 to join Gooch's Regiment. The recruits joined Gooch's Regiment in Jamaica after the Cuba expedition. One wonders what possessed these officers to attempt to lure fellow Americans to serve under the conditions they had seen at Cartagena and Cuba. The following gives an idea of what colonial leaders and potential recruits were being told.

> *"I am informed by Capt. Winslow, who was several weeks in Cuba before his departure thence upon the service of raising recruits here, and is a gentleman well known among us, and of undoubted veracity, that he left 2400 of His Majesty's Forces and 1000 negroes, safely landed in the island of Cuba, on the southerly side of it……where they had seated themselves upon the side of a very commodious, navigable river, in a fertile soil, and found good water and provisions; and so temperate a climate that the soldiers there had quite recovered their health, and were*

in full spirits.....That he (CPT Winslow) *is directed to give forty shillings sterling bounty money out of the money which he is to receive from me, to every volunteer, who shall enlist in the service of the expedition"* **Governor William Shirley to the Massachusetts General Assembly, 23 September 1741, Boston Weekly Newsletter**

"We see also a proclamation to the like purpose by George Clarke Esq.; Lieutenant Governor of New York and an invitation by Capt. William Hopkins of the American Regiment for volunteers to enter, offering grants of land to all such families as are disposed to settle on the island of Cuba." **The Gentleman's Magazine, London, November 1741**

"Newport Rhode Island, Dec. 31, On the 27th inst. The Colony Sloop Tartar, Capt. Wickham Commander, sailed for Cuba, having 53 recruits for Col. Gooch's Regiment." **Pennsylvania Gazette, 10 February 1742**

In February 1742 Captains Farmer, Cosby and Prescott were sent to New York and New England to recruit for Gooch's Regiment, with no recruits found. On 13 April 1742, HMS Stafford departed for Jamaica with 82 recruits for Gooch's Regiment. There is no record of any additional troops being raised in the North American colonies after that.

There is no record of any troops being raised for the other regiments of the expeditionary force. This is perhaps not surprising given the reluctance of men to serve under anyone but their own, known officers shown in the initial raising of Gooch's Regiment in 1740. The recruiting effort continued until disbandment of the Regiment.

> *"As his Majesty has been pleased to put an end to the expedition, and to order me to reduce Col. Gooch's Regiment, and to send them to the northern colonies, I take this first opportunity of signifying to you, that it is his Majesty's pleasure that you put an end to your recruiting, and that you do without delay after this letter shall reach you, discharge such men as you may have raised for the service of the said regiment."* **Major General Wentworth at Kingston Jamaica, to Captain Hopkins, 8 October 1742**

7

Portobello, Ruatan and Charles Town

On 4 February 1742, the HMS Experiment under Captain Dennis departed for Portobello to collect intelligence with thirty three men of Gooch's Regiment on board. Sergeant Duff of Washington's Company was the senior man present and 19 Companies of the Regiment were represented.

During the period of February to May 1742 HMS Rippon sailed with 85 men from the 4th Bn., Gooch's Regiment representing 12 Companies.

During the period of February to July of 1742, the Fire ship Phaeton sailed with 14 men from Captain Innes' Company, Gooch's Regiment.

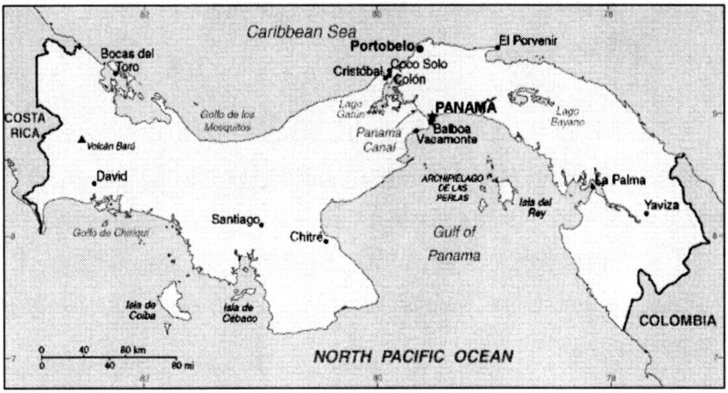

Portobelo, known as Portobello in 1742, is located on the northern most part of Panama

On 9 March 1742, Vice Admiral Ogle sailed from Jamaica, with troops on board under the command of Major General Wentworth, with the intention of landing at Portobello, marching across the Isthmus of Darien, and attacking Panama.

> *"Adm. Vernon is certainly sailed to the Spanish Continent, and it is generally believed he will make an attempt upon Panama. The number of his troops does not exceed 2000 and those sickly, so that the Spaniards need not take arms against us, the climate will do their business. Extract of Letters from Jamaica dated March 14."*

Gentleman's Magazine, London, March 1742

On 28 March 1742 Vice Admiral Vernon's fleet joined Rear Admiral Ogle's at Portobello after having stood off of Cartagena. A detachment of 100 men occupied the Customs House at Portobello the next day. On 30 March 1742, a Council of War agreed that an attempt on Panama was not possible.

> *"Having during their voyage been lessened in their numbers 935 men (officers included) of those that embarked at Jamaica, either by death, or from several transports being separated from the fleet; and from the ships of war being unable to supply them with more that 600 men, being part of the 1,468 embarked on board them, their present numbers would fall short of the 3,000 effective men required for an attempt on Panama."*
> **Major General Wentworth to Vice Admiral Vernon, 31 March 1742**

On 13 April 1742 Vice Admiral Vernon's Fleet departed Portobello for Jamaica, perhaps a wise move given that Spanish sources report that Panama had been reinforced with 1400 soldiers from Peru.

"Extract of a letter from a Gentleman in Jamaica to his friend here, dated Kingston, April 29, 1742. The Expedition is like to end worse than any of the former. Our good Governor is come back, entirely disgusted with both commanders. The best account I can give you of it, is this. Capt. Lowther was sent down to get proper information &c. for a march to Panama from Portobello; and to cover his design, some troops were to be landed both to windward & leeward of that port; the former to take possession of some pass, for securing the march of the main body, which should have been landed at Portobello; by the 28th of February, where they did not arrive till the 26th of March; and the landing to windward had been entirely neglected. The Governor still insisted on landing, which, not being complied with, he borrowed the Greenwich Man of War, and on the 31st left all the rest of the fleet and army at Portobello Harbor, the rainy season being set in……We also hear, that there has been a great mortality among the soldiers and sailors on the expedition."

Pennsylvania Gazette, 24 June 1742
(The governor is Governor Trelawney of Jamaica)

> *"Many have been obliged to submit to the most slavish part of heaving down Men of War in order to be cleaned, a work which does not properly belong to them, besides doing the duty of their own ships at the pump."* **Memorial from the Officers of the American Regiment to the Council of Trade and Plantations, May 1742** (Leach's "Roots of Conflict," p. 59)

On 15 May Vice Admiral Vernon's Fleet arrived at Spanish Town, Jamaica. On 11 July 1742 Lieutenant Colonel Cope died and was not replaced, leaving Gooch's Regiment without a Commander.

> *"Likewise agreeable to the inclinations and views of many officers and men of the American Regiment, he proposed, that a voluntary number of the American Regiment be sent to Rattan, with sufficient provisions for six months, and to be furnished with sufficient arms, ammunition, cannon and all necessary implements, for erection of fortification for*

> *their defense; together with an engineer to direct and assist them; also that Governor Trelawney should give them assurance of a suitable property to every officer and soldier; who should continue in the pay and subsistence of the crown, till they could effectually settle themselves in the lands to be conveyed to them."* **Council of War, Spanish Town Jamaica, 28 June 1742**

During the period of August to October 1742, the HMS Grafton sailed with 100 men from the 4 Virginia Companies of Gooch's Regiment under the command of Captain Washington.

On 13 August 1742, two hundred soldiers of the American regiment, organized into four companies, and fifty marines, were sent from Jamaica, to Ruatan, (currently named Roatan Island) to establish a garrison. Two forts were built, Ft. George and Ft. Frederick.

Roatan Island , known as Ruatan Island in 1742 is located at northern edge of map

On 14 September 1742, the HMS Rippon sailed for Charles Town, South Carolina with 67 men from 4th Bn. Gooch's Regiment. They were part of a contingent of approximately 500 soldiers on 7 transports under the command of Colonel Durour, sent to assist in the defense of South Carolina and Georgia. Vice Governor Bull of

South Carolina and General Oglethorpe of Georgia asked Major General Wentworth to send them following an attack in July on St. Simons Island on the Georgia coast by the Spanish. Finding the Spanish had retreated after their attack in July, Colonel Durour and his detachment left Charles Town for Kingstown, Jamaica on 31 October 1742.

8

Disbanding the Regiment

"The Regiment commanded by Col. Gooch, should be immediately reduced, and that I should send back to North America all such soldiers of that Regiment, as do not choose to remain here in the plantations, or to serve in companies on board the fleet; I have taken the best measures in my power for executing the above orders, and accordingly have dispatched transports to the ports, from whence the soldiers may be most conveniently dispersed to their respective habitations; every man having received his full pay to October the 24th, his firelock, and the clothing of which he was possessed, at the time of his reduction, according to what his majesty was graciously pleased to promise, at the first raising the regiment. As some of the men will probably arrive feeble and sick, I

> *beg leave, sir to recommend it to you, that they may be commodiously conveyed to their respective habitations, as men who have undergone a good deal of hazard, and hardships, during the course of their service in this part of the world."* **Major General Wentworth at Jamaica to Governor Ward of Rhode Island, 15 October 1742**

By late September Major General Wentworth and Vice Admiral Vernon had received orders to return to England. They took with them the two Foot Regiments. Of the six marine regiments raised in England for the expedition, only two remained, and those were divided up between the ships that sailed for England on 19 November from Port Royal Jamaica and the eight Jamaica Independent Companies. On 24 October 1742, Gooch's Regiment was officially "reduced" and the remaining men were returned to their respective colonies over the following months.

> *"By Capt. Howland, from Jamaica, we are informed, that General Wentworth had received Orders from his Majesty to return to Great Britain, with the British Forces under his Command, and to reduce the American Regiment under Col. Gooch,*

> *which he has accordingly done, and sent away all such as refused to take their Discharges at Jamaica, (except a Detachment remaining at Ratan) in four transports, viz. one to New York, two to Virginia, and one to North Carolina. Each Man is paid to the 24th of October last, and allowed to keep his Firelock, and the Clothing he was possessed of at the Time of his Reduction..... It is a very melancholy reflection, that of the five hundred men sent from this province in five vessels at the first embarkation, besides recruits sent at sundry times since,(not to mention two or three hundred sent from Rhode Island) there should not be a sufficient number left to employ one vessel to bring them home. "* **The Pennsylvania Gazette, 4 January, 1743**

On 9 November 1742 four ships departed Jamaica with 346 men, the last of Gooch's Regiment that returned to North America. There may have been other ships bringing soldiers home. In January 1743, Captain Lawrence Washington returned to Virginia. Approximately 4,000 men had served in Gooch Regiment since its inception.

> *"By a Letter from North Carolina we hear, that the Mary Transport is arrived at Wilmington in that Colony from Jamaica, with only 25 Men, reduced from 100; which Letter farther adds, that a large Fort was building on the Island Ratan, in order to secure the American Trade to the Mosquito Shore and the Bay of Honduras, and that 300 Men were enlisted to garrison the same."* **Caledonian Mercury, Edinburgh, May 2, 1743**

In 1743, Lieutenant Brown from Gooch's Regiment and supposedly other veterans of Gooch's Regiment, were among the 200 Virginians recruited by Major Alexander Heron to support General Oglethorpe in Georgia. On 25 December 1743, 8 Independent Companies of Jamaican Volunteers and the remnants of Gooch's Regiment on the Island of Ruatan were formed into Trelawney's 63rd Regiment of Foot, with the red coats and green facings and cuffs worn by Gooch's Regiment and the Independent Companies in North America.

9

Conclusions

> *"From the highest to the lowest, the inhabitants of these provinces seem to set great value on themselves, and think a regard is due to them, especially in the assistance they are able to give the Mother Country on such occasions; and as they are a growing power, should they be disappointed in what is promised them and which they expect, future occasions of the like nature may suffer for it."* **Colonel Blakeneys' Letter to the Council of Trade and Plantations, May 1741** (Leach's "Roots of Conflict," p. 51)

Gooch's Regiment was raised, at least in the view of those who would support it, recruit for it and serve in it, to provide a land force element to participate in British expeditionary operations in the Caribbean. Gooch's Regiment has no lineage in the British Army. Its numerical designation, the 43rd, was given to another regiment.

The Royal Marines and United States Marine Corps claim it as a predecessor and rightly so, as its members were referred to repeatedly as marines. The claim is no empty boast, as the Regiment did have opportunity to participate in operations characteristic of the 18th century marine, those being typically limited to shipboard gun crews, sharpshooters in the rigging, boarding parties and landing parties. Many in the regiment served on ships as replacements for common seamen, counter to their understanding of what they enlisted for. The conditions under which they served were especially brutal, even for the time, with Officers uninformed of their men's whereabouts and housed on ships where they had no place to sleep. They were also removed from the Naval Hospital in Jamaica counter to the regulations governing the medical treatment of marines serving on ships.

The Crown was not prepared for the number of recruits raised for the regiment. That is apparent for the 15 companies raised over and above the 30 planned that never served in Gooch's Regiment. These 1200 men, given the choice of believing the promise of commissions for their American commanders and proper arms and equipment or disbandment, chose disbandment. The Crown underestimated the willingness of the colonies to support the raising and transporting of the Regiment, as well as to loan the money for their pay, in order to participate in an expedition that offered no material advantage to them. It is quite possible that had Cartagena, Portobello and or Santiago been captured that it might well have damaged the financial interests of those maritime colonies involved in a brisk smuggling trade with the Spanish ports in the Caribbean.

From the outset, however, it became apparent that the senior leaders of the expedition had every intention to use the Regiment primarily as a labor force for the land forces and a source of manpower to replace sailors in the fleet. The American soldiers were seen by the British as untrained, but there is no record of any effort to train the new soldiers, NCOs, or Officers after they arrived in Jamaica in 1740. Being untrained made them more expendable, of far less value than their trained European counterparts and useful as a labor pool for the land and sea forces. Though not stated by contemporary sources, it is easy to get the impression that the British thought that the soldiers raised in North America would be acclimatized to the Caribbean. This was obviously not the case, and, despite the casualties, the British, at least throughout the Cartagena expedition, continued to act as though it were true, causing enormous loss of life in Gooch's Regiment.

Colonists back in North America were aware of how the men of Gooch's Regiment were treated and the loss of life, perhaps as much as 80% of the Regiment, was bound to have made an impression as well. It is no small wonder that the British found it difficult to raise regiments to serve under European Officers and NCOs in the years to come.

Sources

Beatson, Robert, "Naval and Military Memoirs of Great Britain, from 1727 to 1783' Vol. I., Longman Hurst, London, 1804

Berkeley, Francis, "The War of Jenkin's Ear", in " The Old Dominion, Essays for Thomas Perkins Abernethy," University Press of Virginia, Charlottesville, 1964

Boyse, Samuel, "An Historical Review of the Transactions of Europe," Vol. I., D Henry, Reading, 1747

Browne, William, Ed., "Proceedings of the Council of Maryland, 1732-1753," Maryland Historical Society, Baltimore, 1908

Carman, W.Y., "The Royal Artillery," Osprey Publishing, New York, 1973

Clark, Mertie J., "Colonial Soldiers of the South, 1732-1744," Genealogical Publishing Co. Baltimore, 1986

Clark, Walter, ed., "The State Records of North Carolina," Vol. XV, Nash Brothers, Goldsboro, 1898

Dalton, Charles, "English Army Lists and Commission Registers, 1661-1714." Vol. V., Eyre and Spotiswoode, London, 1902

Foote, William, "The American Independent Companies of the British Army, 1664-1764", University of California Ph. D. Thesis, 1966

Ford, Douglas, "Admiral Vernon and the Navy, A Memoir and Vindication," Fisher, London, 1907

Fortescue, J.W., "A History of the British Army," Vol. II, MacMillan and Co., New York, 1899

Gardner, W.J., "A History of Jamaica," Elliot Stock, London, 1873

Hall, Clayton, " Maryland's Part in the Expedition against Cartagena," Maryland Historical Magazine, Vol. II, pp. 146-159

Harding, R.H., "Amphibious Warfare in the Eighteenth Century, The British Expedition to the West Indies, 1740-1742," The Boydell Press, 1991, Suffolk, UK

Harding, R.H., "The Growth of Anglo-American Alienation: The Case of the American Regiment, 1740-42," The Journal of Imperial and Commonwealth History, Vol 17, No. 2, 1989, pp. 161-184

Harkness, Albert, "Americanism and Jenk's Ear," The Mississippi Valley Historical Review, Vol. 37, No. 1 (Jun 1950), pp. 61-90

Hazard, Samuel, "Pennsylvania Archives," Vol. I, Joseph Severns, Pittsburg, 1852

Hening, William Walter, "Hening's Statutes at Large, Being a Collection of all the Laws of Virginia from the first session of the Legislature, in the Year 1619." http://www.vagenweb.org/hening/

Hewins, W.A.S., "The Whitefoord Papers," Clarendon Press, Oxford, 1898

Jones, Alfred, "The American Regiment in the Carthagena Expedition," Vol. XXX, The Virginia Magazine of History and Biography, Jan 1922

Jones, Charles, "The History of Georgia," Vol. I, Houghton, Mifflin and Co., Boston, 1883

Kimball, Gertrude, "The Colonial Governors of Rhode Island," Vol. I, Houghton, Mifflin and Co., Boston, 1902

Knowles, Charles, "An Account of the expedition to Carthagena: with explanatory notes and observations," M. Cooper, London, 1743

Leach, Douglas E., "Roots of Conflict, British Armed Forces and Colonial Americans, 1677-1763", University of North Carolina Press, Chapel Hill, 1986

McBarron, Charles, et al., "The American Regiment, 1740-1746," Military Collector and Historian, Vol. 21. No. 3 (Fall 1969), pp. 84-86

Nicholas, Paul, "The Historical Record of the Royal Marine Forces," Vol. I, Thomas and William Boone, London, 1845

Ogelsby, J.C.M., "England vs. Spain in America, 1739-1748:The Spanish side of the Hill, " Historical Papers, Vol V, No. 1. 1970, pp. 147-157

Porter, Whitworth, "History of the Corps of Royal Engineers," Vol. I, Longman, Green and Co., London, 1889

Pawson, Michael, et. al., "Port Royal Jamaica," University of the West Indies Press, Kingston, 2000

Rea, Robert, "Major Robert Farmar of Mobile," University of Alabama Press, Tuscaloosa, 1990

Rolt, Richard, "An Impartial Representation of the Conduct of the Several European Powers Engaged in the Late General War." Vol. II, T. Longman, London, 1749

Rutherford, Livingston, "Family Records and Events complied principally from the Original Manuscripts in the Rutherford Collection," Devine Press, New York, 1894

Smollett, Tobias, "An Account of the Expedition Against Carthagena in the West Indies, Besieged by the English in the Year 1741," The Complete Works of Tobias Smollett, Vol. II, Bell and Daldy, London, 1869

Smollett, Tobias, "Roderick Random," Routledge and Sons, London, 1895

Stiles, Henry, "The History of Ancient Windsor Connecticut," Charles Norton, New York, 1859

Syrett, David, "The Raising of American Troops for Service in the West Indies during the War of Austrian Succession, 1740-1," Historical Research, Vol. 73, No. 180 (February 2000), PP. 20-32

Vernon, Edward, "Original Papers Relating to the Island of Cuba," M. Cooper, London, 1744

Vernon, Edward, "Authentic Papers Relating to the Expedition Against Carthagena," L. Raymond, London, 1744

Collections of the Massachusetts Historical Society, The Belcher Papers, Part II, Sixth Series. Vol.. 54, Massachusetts Historical Society, Boston, 1894.

"American and West Indies: December 1739". Calendar of State Papers Colonial, American and West Indies, Volume 45: 1739(1994), pp. 234-273

Anon., "The Gentleman's Magazine," April, 1740, London

Anon., "The Gentleman's Magazine," February, 1741, London

Anon., "The Gentleman's Magazine," May, 1741, London

Anon., "The Gentleman's Magazine," November, 1741, London
Anon., "The Gentleman's Magazine," February, 1742, London
Anon., "The Gentleman's Magazine," March, 1742, London

Anon., "The Cloathing Book 1742," The Naval & Military Press, LTD. East Sussex, England.

Country and Regional Maps from the CIA World-Fact Book: https://www.cia.gov/library/publications/the-world-factbook

About the Author

Lee Offen is a retired U.S. Army Major who resides in Springfield Virginia. He is an Armored Cavalry Combat Veteran, former Airborne Infantryman and a recipient of the Bronze Star.

He has his Masters Degree in Government and was Assistant Professor of Military Science at Georgetown University and George Mason University. A graduate of the U.S. Army Military History Instructors Course he is also a historical re-enactor and avid student of 17th and 18th century American and European history.

Breinigsville, PA USA
07 October 2009

225407BV00001B/4/P